Astonishing Book

by Linda Strachan
illustrated by Debbie Mourtzios

EDUCATORS PUBLISHING SERVICE
Cambridge and Toronto

Series Authors: Kay Kovalevs and Alison Dewsbury
Series Consultant: Melinda Rice
Commissioning Editors: Sarah Russell, Bonnie Lass, and Sethany Rancier Alongi
Senior Editorial Manager: Sheila Neylon
Text by Linda Strachan
Illustrated by Debbie Mourtzios
Edited by Janet Mau, Elissa Gershowitz, Stacey Nichols Kim, and Laura A. Woollett
Designed by Colby Heppéll

Making Connections program developed by Educators Publishing Service, a division of School Specialty Publishing and by Harcourt Education, a division of Reed International Books Australia Pty Ltd

ISBN 0-8388-3335-7

2010 2009 2008 2007 2006
10 9 8 7 6 5 4 3 2 1

Printed in China

Contents

CHAPTER ONE
The Attic
page 5

CHAPTER TWO
A Carved Wooden Box
page 10

CHAPTER THREE
The Book
page 16

CHAPTER FOUR
Awkward Questions
page 22

CHAPTER FIVE
What Does Jennifer Know?
page 27

CHAPTER SIX
At the Historical Society
page 31

CHAPTER SEVEN
Jennifer's Treasure
page 37

CHAPTER ONE

The Attic

The attic door creaked as Trish pushed it open. It was dark inside, and she had forgotten to bring a flashlight. Trish hated the dark, dust, and smell of attics. She had only agreed to go up there with her twin sister, Jo, because there was nothing else to do—except help their stepmother, Jennifer, get settled into their new house.

The girls were tired of unpacking. There were too many heavy boxes of books, most of them Jennifer's. Jennifer was an artist who restored old books. People brought her their books and paid her to clean them up and rebind them so they looked like new. Trish and Jo thought Jennifer's job was really cool until they had to lug all the heavy boxes of books around and help her unpack them.

Earlier in the day, Jennifer had asked if they could help her around the house. She was always trying to be their friend.

"Do you want to help me choose a color to paint the dining room?" Jennifer asked. "I'm going to pick up the book of colors at the store later."

"Yeah, whatever," Trish sighed.

It wasn't that Trish didn't like Jennifer. She just didn't know her very well yet. It had only been Trish, Jo, and Dad for such a long time. It was kind of weird to have someone else around. And Jennifer worked from home, which meant she was around a lot.

The house was really Jennifer's idea. She had grown up in this town and remembered the house from her childhood. A friend of hers, Maggie Dalton, had lived in the house with her parents and brothers.

When they were girls, Jennifer and Maggie spent hours playing together in the big backyard and under the eves of the attic. Jennifer had fond memories of the house.

The Daltons had moved away a long time ago, and other families had moved in and out over the years. Jennifer always kept her eye on the house to see when it came up for sale. This time, she and Dad had decided to buy it, so here they all were.

Jennifer and Dad were excited about the new house. However, Trish and Jo were sad about leaving all their friends and worried about making new ones. There were a lot of changes all at once. At least the girls knew they had each other.

CHAPTER TWO

A Carved Wooden Box

Standing in the middle of the attic, Trish and Jo looked around. It wasn't as dark as it had seemed at first. At the far end of the attic was a small window. It was covered in dust and grime, but some of the summer sunlight shone through.

"This is astonishing!" Jo gasped as she followed her sister toward the light. *Astonishing* was Jo's new favorite word. "I never thought the attic would be so big."

Trish agreed. "But look at all this junk. And it's so dusty. I don't think anyone's been up here for years. I'll bet this stuff has been left behind by all the people who've ever owned this house."

The girls began exploring. As Trish climbed over old rolls of carpet to get to a bike that was leaning against the wall, Jo started looking through some boxes. Most of them contained old papers, old clothes, and junk. Jo pushed one box aside to get to the next and froze as something unusual caught her eye.

There in the corner was a carved wooden box with a hinged lid and a metal handle

on each side. The box was heavy, and the lid wasn't quite straight. It obviously hadn't been opened in a long time because the hinges looked like they had rusted shut.

Jo dragged it closer to the window, to get a better look. "Trish, come help me," she called urgently, her voice filled with excitement. "I can't get the top off this box. And it looks like this one might actually be interesting!"

The girls knelt in front of the box. Jo tugged on the top, gently rocking it from side to side to loosen the rusty hinges. Slowly, the box groaned open, and the girls eagerly looked inside. Soft silky material covered the object within.

Jo reached in and pushed aside the material. She gasped in amazement as she saw the cover of a large handmade book.

CHAPTER THREE

The Book

Jo carefully lifted the large book out of the box. The curved lettering of the title looked like it was written in a foreign language. The cover was decorated with beautiful illustrations of leaves and flowers wound together in rich colors. The girls looked at the book in wonder.

"Wait, Jo, our hands—and everything around us—are filthy," exclaimed Trish. "I'll run downstairs to get our white gloves and a towel to put the book down on before we open it."

Jennifer had explained to Trish and Jo that the natural acids on people's hands can damage old books, which is why she always wore white cotton gloves when handling them. Jennifer had given Trish and Jo each their own pair. At last, they had the chance to use them.

Trish returned with the white gloves and towel. Jo started turning the pages, examining what was on each one. She stopped when she came to a beautifully painted illustration. "Wow, look how fancy it is, Trish."

Trish sneezed. "It's very beautiful. But can we take it somewhere else to look at it? My nose is itchy from all this dust," she complained, "and I'm hungry."

"Well, I'm going to do some research on the writing on the cover. I think the book is astonishing!" announced Jo, covering the book with the silk wrapping again and placing it back inside the box.

"Let's go to the library after lunch," suggested Trish. "It's better than staying around here and unpacking."

"Okay," agreed Jo. They walked down the stairs together. "In the meantime, let's not tell anyone about what we found. Once we figure out what it is, we can surprise Dad and Jennifer!"

"Surprise Dad and Jennifer with what?" asked their stepmother, as the girls walked into the kitchen. "What have you two been up to?"

"Just looking around the attic," replied Jo, sitting at the table.

"The attic?" Jennifer repeated. "Did you find anything interesting up there?"

"Nah," answered Jo, sneaking a look at her sister. "Just old junk. Lots of boxes. Rolls of dusty carpet."

"I used to love attics when I was a child," Jennifer said softly, sitting down next to the girls. "Especially the one in this house. My friend Maggie and I spent hours upstairs, when she lived here . . . " her voice drifted off and she gazed into the distance.

The two girls looked at each other. After a moment or two of silence, Jennifer sighed and stood up. "Are you hungry?" she asked. "It's probably lunchtime."

After lunch, Trish and Jo slipped back up to the attic to get the book. They rushed to their room, grabbed a backpack, and then started outside to get their bikes. As Trish was about to slam the door behind them, Jennifer poked her head around the corner. "Don't be late for dinner," she reminded them. "Your dad is making your favorite meal—spaghetti and meatballs."

"Yeah, okay," Trish replied impatiently.

CHAPTER FOUR

Awkward Questions

Jo and Trish made the five-minute trip to the library in silence. Jo was thinking about the book, and Trish was thinking about how she already missed life in their old house. Trish wished that she, Jo, and their dad were still living there, just the three of them, like old times.

When Trish and Jo got to the library, Jo spent the afternoon searching for information on the strange writing in the book. She discovered that it looked a little bit like old English writing, but it wasn't quite the same.

Meanwhile, Trish looked at all the pages of the book carefully, comparing them to the examples Jo found. "I think we should take the book to the historical society," she suggested. "Someone there might recognize the writing."

Jo agreed. "But the bike path doesn't go that far. Maybe Dad can drive us there tomorrow."

When the girls arrived home, Jennifer was sitting at the kitchen table. Her clothes were covered in dust. Trish guessed, "You've been in the attic!"

Jennifer replied, "Yes, I was trying to make some space so we can store our things. You were right, girls. There is a lot of trash up there. But I wanted to ask you something."

Trish glanced at Jo, but Jo was avoiding her eyes.

"I saw a box on the floor with some rusty hinges. It looks like someone forced off the top, and I was hoping you girls didn't hurt yourselves." She stopped, waiting for Trish and Jo to answer.

"No, it came off pretty easily," said Jo. "Too bad the box was empty."

Jennifer wasn't convinced. She asked, "Are you sure the box was totally empty?"

"Totally!" Jo and Trish said together.

Jennifer looked at them strangely, but she didn't say anything else.

CHAPTER FIVE

What Does Jennifer Know?

Dad came in with the groceries and began preparing dinner. The kitchen filled with the wonderful smells of fresh herbs and rich spaghetti sauce. The girls were soon laughing

and talking as they helped make the garlic bread and prepare the salad.

Jennifer came downstairs and the family sat down to dinner. The girls ate in silence while their father and Jennifer chatted.

As soon as Trish and Jo had finished eating, Trish asked, "May we be excused?"

“Of course,” replied Dad, “but don’t you want some dessert?”

“No thanks. We’ve got homework to do,” said Trish.

“Oh, yeah!” Jo blurted out. “Dad, can you drive us to the historical society after school tomorrow?”

"Sure!" said their father, brightening. "What do you want to do at the historical society?"

Jo had to think of something quickly. "Oh, just some research for a school project," she said evasively.

"Okay," said Dad. "I'll pick you up after school."

As the sisters went upstairs, Trish wondered, "Do you think Jennifer knows something about the book?"

"What if . . . " Jo was thinking hard. "What if she already knew the book was in that box in the attic and she couldn't get there in time to look for it? Maybe we beat her to it."

Trish sighed. "Yeah, I don't think she believed us when we said the box was empty. What are we going to do?"

"Well, we already have a plan, remember? We're taking the book to the historical society tomorrow," Jo replied. "And won't it be fun to show her everything we find out!"

CHAPTER SIX

At the Historical Society

The next day, Dad drove the girls to the historical society. At the front desk, they asked to see the antique books expert. A few minutes later, the girls were in her office.

"Hello, I'm Marisa. What can I do for you?" she asked.

Jo took the book carefully out of her backpack and unwrapped it. She explained where she and Trish had found it and how they had tried to research the writing but had found out very little about it.

Marisa put on a pair of white cotton gloves and Jo handed her the book. "I'm putting on gloves," she said, "because the natural acids on our hands can ruin old objects."

"Yes, we know!" Trish and Jo said together. Trish continued, "We wore white gloves when we touched the book because we didn't want to damage it."

"Good job!" exclaimed Marisa. "How did you know to do that?"

"Our stepmom restores old books," explained Jo, peering over at the book.

Marisa began turning the pages of the book on her desk.

"I did some research yesterday and I think it's an illuminated manuscript," Jo said.

"What's an illuminated manuscript?" asked Trish.

"It's a book that's specially decorated with borders and pictures all painted in bright colors," Jo replied.

"Well, I hate to have to disappoint you both, but this isn't an authentic illuminated

manuscript," said Marisa. "From the kind of paper used, I can tell the book isn't very old at all."

"Doesn't this look like one of the books we helped Jennifer unpack in her studio the

other day? Maybe Jennifer made the book," Trish wondered. "What do you think, Jo?"

Jo nodded, "I bet you're right!"

"Well, the mystery is solved," Marisa smiled. "Come back any time and bring Jennifer with you. We have a large collection of illuminated manuscripts that I'm sure she will find interesting."

"Let's go home and share our discovery with Jennifer!" said Jo. "If she did make the book, she's going to be even more excited to see it than we thought."

CHAPTER SEVEN

Jennifer's Treasure

When Dad and the girls arrived home, they found Jennifer in the kitchen, sitting at the table. Her clothes were covered in dust again.

"How did your research for your school project go?" Jennifer asked.

"It was very 'illuminating'," replied Jo.

Jennifer looked confused.

After groaning at the corny joke, Trish said, "We have something special to show you, Jennifer."

"You do?" Jennifer asked, surprised.

Trish and Jo handed the book to Jennifer, who gasped when she saw it.

"I didn't dare hope that any of them had survived," she exclaimed. "Oh thank you,

Trish! Thank you, Jo! When we were young, my friend Maggie and I used to make books that looked like illuminated manuscripts. At that time Maggie lived in this house. We spent hours in the attic making the books. I lost touch with Maggie and her family, so I never knew what happened to our creations. I'm so happy to have one of them back!"

Jennifer put her arms around the girls and hugged them tightly. Then they looked through the beautiful pages of Jennifer's newly recovered treasure together.

"I wonder if any more of our creations are up in the attic," Jennifer said. "Shall we check out the attic this weekend, girls?"

"Yes!" Trish and Jo replied together. "And maybe you could show us how to make our own books?" Trish asked shyly.

"I would love to!" Jennifer smiled.

"There's so much detail in the illustrations, Jennifer. It must have taken you and Maggie a long time to create this book. It's astonishing!" exclaimed Dad.

"Hey, that's my word!" laughed Jo.